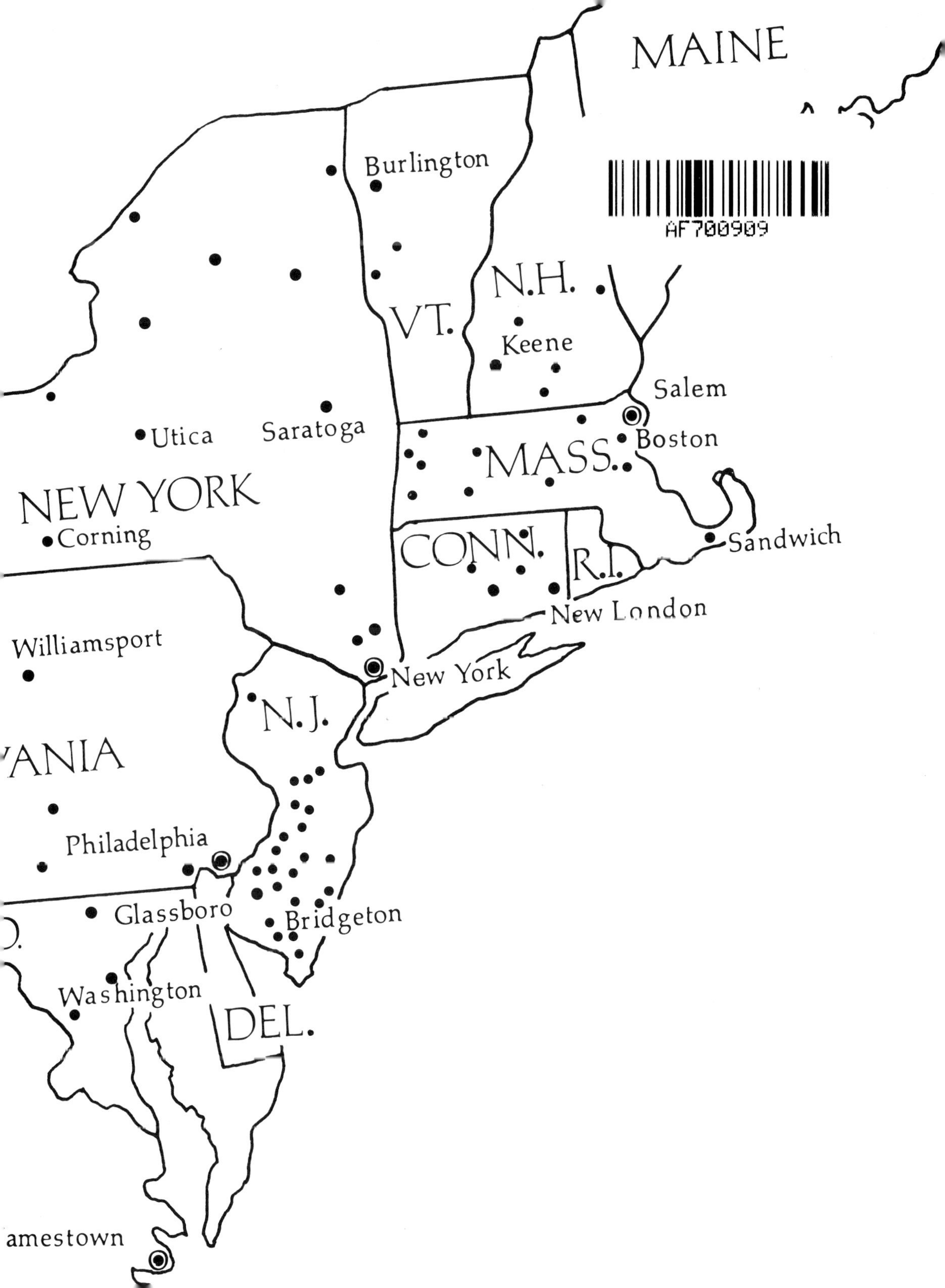
MAINE
Burlington
VT.
N.H.
Keene
Salem
Boston
MASS.
Utica
Saratoga
NEW YORK
Corning
CONN.
R.I.
Sandwich
New London
Williamsport
New York
N.J.
Philadelphia
Glassboro
Bridgeton
Washington
DEL.

Fig. 2. *Decanter made at Cha*
Ihmsen's glasshouse, Pittsburgh,
1813, engraved with a view of
first naval victory of the War
1812. (55.4.44).

GLASSMAKING

AMERICA'S FIRST INDUSTRY

Jane Shadel Spillman

The Corning Museum of Glass

CORNING GLASS CENTER · CORNING, NEW YORK 14830

PRINTED IN U.S.A.

Standard Book Number 0-87290-062-2
Library of Congress Catalog Card Number 76–17354

INTRODUCTION

Glassmaking was America's first industry, but the colonies had been settled over a hundred years before a successful glass factory was well-established. The English laws governing the colonies included restrictions prohibiting manufacture. As Americans struggled for political independence so also did they pursue economic independence, and this full economic independence was finally achieved in the nineteenth century when American manufacturers were successful in competing with Europeans. American glassmakers captured a larger and larger share of the home market and began to export glass both to Europe and to other developing countries. By the late nineteenth century, the United States was a world leader in the manufacture of glass and has remained so in the twentieth century. American glassware competes successfully all over the world and America makes a large share of the glass products produced today. This is a far cry from the six men who came to the New World in 1608 to begin the manufacture of glass.

3

4

Fig. 3. *Casement window with small panes of the type used in the 17th century.*

Fig. 4. *Wineglass, English, ca. 1700. (54.2.11).*

Fig. 5. *Spirits bottles, English, 18th century. (60.2.11; 60.2.14; 60.2.5).*

5

FIRST INDUSTRY IN AMERICA

When the Jamestown colony was established, the settlers brought a group of glassblowers with them. The English Board of Trade decided that glass production would be an ideal use for the limitless forests of the New World. England at this time was being rapidly denuded of wood and by 1615 a Royal Proclamation to preserve the remaining timber for building ships of the Royal Navy left the glassmakers no choice but to learn to use coal in their furnaces. England was still struggling to establish its own glass industry—most Englishmen, when they could afford glass, drank from imported Venetian wares.

A lack of technical knowledge on the part of the English financiers, the terrible hardships of the New World, the quarrelsome natures of the glassblowers sent to Jamestown, and the sheer distance to the markets in England all combined to make the first attempt at glass production in the colonies a failure, in spite of the abundance of fuel and good sand, the chief ingredient of glass. Glass factories were started in Salem, Massachusetts, in New Amsterdam, in Philadelphia, in Germantown, Massachusetts, and other places, but all were failures. Throughout the seventeenth and early eighteenth centuries the colonists were forced to import glass for their windows, their wine and their tableware. In 1629, the Rev. Francis Higgins wrote from Salem, Massachusetts to friends in England, "Be sure to furnish yourself with glass for Windows," and storekeepers' advertisements of the eighteenth century offered "Looking-Glasses, Double Flint Decanters, single Flint Wine Glasses by the Dozen . . . all sorts of Glass Bottles . . . Case and Pickle Bottles from one quart to a gallon; quart and pint Black Bottles by the Gross, Glasses for Chymists, and the Window Glasses of different sizes."

6

Fig. 6. *Glass probably made at one of Stiegel's glasshouses, Pennsylvania, ca. 1763–1774.*

Fig. 7. *Sugarbowl, glasshouse of Caspar Wistar, Wistarburgh, New Jersey, ca. 1739–1781. (50.4.2).*

INDEPENDENCE PURSUED

The first successful glass manufactory in the colonies was that of Caspar Wistar, a Philadelphia brass-button manufacturer, who had immigrated from Germany. He imported German glassblowers in 1739 to staff the factory he established in southern New Jersey. Official English policy at this time forbade all manufactures in the colonies: America was to produce raw materials for factories in England and provide a market for the finished goods. Wistar defied this ruling to produce windows and bottles, commodities then much in demand. One eighteenth century newspaper reported that:

> "No factory is more wanted . . . . Than a capital glass manufactory for bottles similar to those of Bristol in Great Britain. Considerable orders for Philadelphia beer and porter are now in the city, some of which cannot be executed for want of proper bottles. . . ."

The second successful entrepreneur in glass was also a German. Henry William Stiegel built three glasshouses at Elizabeth Furnace and Manheim, Pennsylvania between 1763 and 1774, and attempted to produce fine tablewares as well as bottles and window glass. Stiegel also imported workmen; there was no other way to get them and, like Wistar, he had problems keeping his workmen, some of whom wished to organize their own factories. The newspaper advertisements of both Wistar and Stiegel appealed unsuccessfully to the patriotism of their fellow citizens to buy American made rather than imported wares. Stiegel so overextended himself financially with his three factories that he went eventually to debtors' prison in 1774; but Wistar, who was perhaps a better businessman, left a successful factory to his son who closed it in 1781, probably because of the Revolution.

Fig. 8. *"Tobias and the Angel" tumbler made at Amelung's New Bremen Glassmanufactory for his wife, Carolina Lucia Amelung, dated 1788. (55.4.37).*

AMELUNG AND THE QUEST FOR EXCELLENCE

John Frederick Amelung arrived in the new republic in 1784 with men and equipment to develop a large glass factory complex for the manufacture of all types of glass. He came from a glassmaking family in Germany and was backed by a group of merchants in Bremen.

Amelung's factory was established in Maryland, at a site he named New Bremen and was, for a time, successful. He produced the most sophisticated glass which had been made in America up to that time. But he was unsuccessful at getting the new Congress to pass protective tariffs and he was undersold by his English competitors.

A contemporary view of the situation perhaps best describes why he and others after him failed.

> "Most new works have been begun too large in this country. . . . If we built a glasshouse, it was at the expense of Thousands and calculated to cover all that part of the country with glass which was not covered by houses."

Nonetheless, Amelung left us the best quality wares made until that time in America; his was the first signed and dated glass, so that we may reliably identify his work. His factory burned in 1790, was rebuilt, and finally closed in 1795.

INDEPENDENCE ACHIEVED—1800

After the Revolutionary War there were no more restrictions against manufacture, but British mercantile policies made it difficult for American manufacturers to compete. Finally, after 1790, Britain signed trade agreements with the new country; when the Acts of Embargo and the War of 1812 cut off the supply of British-made goods, Americans for the first time were free of ruinous competition. Sixty-three glass factories were started in the new republic between 1790 and 1820, but thirty-four of them failed when imported goods once more flooded the market after the war. In spite of this, glass production increased during every decade throughout the nineteenth century. The first successful factory in New England was the Boston Crown Glass Manufactory, organized in 1788 but not in production until 1793. This factory made window glass which was the equal of any imported, and bottles and table glass as well. Most of the factories started in those first decades were primarily for bottles and/or window glass; such tablewares as were made were usually made by individual workers for family or friends. Most of the early nineteenth century window glass factories in New Jersey, New England and New York permitted their workmen to make table glass on their own time at the end of a shift and this "off hand" glass, often decorated with superimposed gathers and threading in a Germanic tradition, is as close to a folk art form as the glass industry produced. No two pieces of this glass are ever exactly alike since design depended on the whim and skill of the workmen and all kinds of household articles were made. The growth of transportation systems, such as the National Road which led west from Baltimore, and the Erie Canal, facilitated the spread of glass factories and availability of their products. Water transportation was especially important as it was cheaper and less apt to damage the goods. Packets plied the East Coast selling New England's glass, which was also shipped to the West Indies.

It is not surprising that so many factories were started in this period when the future of glass manufacture seemed so hopeful. One European traveler reports,

Fig. 9. *Sugarbowl made at the South Boston glasshouse of Thomas Cains, ca. 1813. (73.4.116).*

Fig. 10. *Advertisement of the Boston Crown Glass Manufactory from a billhead, 1821. The glasshouse and its products are shown in this advertisement.*

Fig. 11. *Group of objects blown at window and bottle glasshouses in New England, New Jersey and New York, ca. 1820–1850.*

10

11

". . . . it is only necessary to reflect upon the situation of the United States. They have immense forests to clear, consequently it is highly proper that they should establish glass manufactories, and increase them as much as possible. The labour employed to destroy the wood for the clearing of lands, at the same time that it disposes the land to culture, will serve for the production of a very extensive object of manufacture, therefore the utility of this destruction is double to the Americans. It cannot be doubted, that this consideration will strike them, that they will one day conceive the project of furnishing Europe with glass-ware. . . ."

MOVEMENT WEST

As settlers moved west to the Alleghenies, they created a new market for manufactured goods, but glass proved difficult to ship overland and western settlers had to do without it in their windows and on their tables unless they were very wealthy. A traveler to the Western Reserve in the early nineteenth century commented, "The furniture for the table is equally scanty and inconvenient . . . articles of crockery are few and indifferent. . . . For want of a glass . . . from which to drink, if you are offered whisky (which is the principal drink here) the bottle is presented to you or a bowl or a teacup containing the liquor. . . ."

This situation began to improve in the 1790's when James O'Hara and Isaac Craig of Pittsburgh started a bottle manufactory—luring glassworkers from Philadelphia. A second factory was started south of Pittsburgh by Albert Gallatin, later Secretary of the Treasury, employing glassblowers who had previously worked for Amelung.

Pittsburgh was an ideal location for manufacture of all kinds, because river transportation to the entire western frontier guaranteed a ready market and nearby coal deposits provided readily available fuel. Other factories were started and by 1817, when President Monroe wanted cut glass for the White House, he ordered it from Benjamin Bakewell's Pittsburgh glasshouse. Bakewell's was the largest factory in Pittsburgh for many years, shipping glass East and West and to Mexico and the Indies. From Pittsburgh, the glass industry spread down the Ohio to western Virginia and the Western Reserve after the War of 1812. Twelve factories were started in western Virginia, Kentucky and Ohio before 1825.

12

Fig. 12. *Tumblers with cut decoration and enclosed sulphide portraits of Jackson and Washington, made at Benjamin Bakewell's Pittsburgh glasshouse, ca. 1824. (55.4.57; 55.4.273).*

Fig. 13. *Group of objects made in Ohio glasshouses, ca. 1815–1840.*

13

Fig. 14. *Group of blown three-mold glass made in New England and Ohio glasshouses, ca. 1815–1830.*

Fig. 15. *Group of American liquor flasks, ca. 1820–1870.*

15

MASS PRODUCTION

The manufacture and decoration of hand-blown tableware is slow and costly and it was not long before manufacturers were seeking ways to speed production and decorate their glass more cheaply. One way to do this was to blow the glass into a mold, producing shape and glass surface pattern in one operation. The earliest of this "moulded" glass, as it was called, imitated cut glass. A housewife's book, published in New York in 1815, suggests,

> "Those who wish for Trifle dishes, butter stands, &c. at a lower charge than cut glass may buy them in moulds, of which there is a great variety that looks extremely well if not placed near the more beautiful article."

This type of glassware, called blown three-mold by collectors, was popular for about twenty years until it was superseded by pressed glass.

It was always difficult to find skilled glassblowers as they were continually on the move, seeking better wages or the free land available in the western territories. Labor problems probably contributed to the eagerness of manufacturers to use molds. Bottles and flasks for liquor were also blown in full-size molds and after the War of 1812 were made with decorative patterns of all types including portraits of political heroes, Masonic symbols, trade symbols, landmarks and the American eagle. George Washington's face appears on sixty-one different types of flasks, made from New Hampshire west to Pittsburgh, and Benjamin Franklin, Zachary Taylor, Lafayette, DeWitt Clinton and Andrew Jackson are only a handful of the other politicians so honored. The grain crops of the western farmers were most easily transported to market after conversion into whisky—corn was thus a cash crop and the thousands of figured flasks produced attest to its importance.

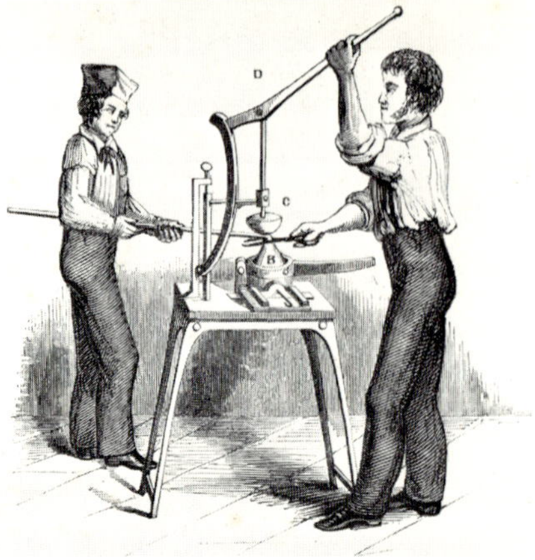

16

PRESSING

The use of molds for blowing certainly speeded production, but it was an idea borrowed from England. The pressing machine developed in the 1820's tripled American tableware production and, as one Englishman said in 1829, "The merit of its invention is due to the Americans and it is likely to prove of great national importance."

Operation of the pressing machine took two men instead of three or four: one to bring the gather of hot glass from the furnace and drop it into the machine, and the second to pull the plunger and operate the press. Cutting off the exact amount of glass needed took skill, but not nearly so much as the blowing of an object. A bowl came out of the press completely decorated in a few seconds. Because contact with the cold mold produced an ugly network of chill wrinkles on the surface of some pieces, moldmakers turned from imitations of cut glass patterns to decorative patterns with stippled backgrounds—where every inch of the piece was decorated the wrinkles didn't show. Technological improvements finally eliminated the chill mark and patterns became simpler in the 1840's.

Objects like candlesticks, vases and lamps could be produced in separate parts and fused while hot, making it possible to produce more glass from fewer molds. Finally, table glass was produced in sets like china; for the first time housewives could set a table with matching glass, which naturally encouraged them to buy more. The invention of the pressing machine was the single most important contribution of the American industry to glass technology.

Fig. 16. *Pressing machine, from a contemporary illustration.*

Fig. 17. *This variety of lighting fixtures was produced by combining a few basic pressed parts into a number of styles. New England, ca. 1840–1860.*

Fig. 18. *"Comet" pattern glass, made in both New England and the Pittsburgh area, ca. 1850. (68.4.12-23).*

Fig. 19. *Group of "lacy" pressed glass, New England, ca. 1827–1840.*

17

18

19

Fig. 20. *Whale oil lamps, New England, ca. 1818–1840.*

Fig. 21. *Replica of the Westinghouse bulb used at the Columbian Exposition, Chicago, 1893. (72.4.87).*

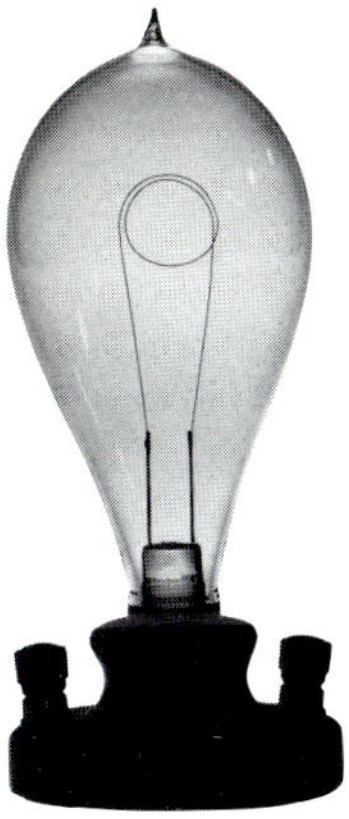
21

GLASS IN LIGHTING

Before 1800, most of the lamps and candlesticks used in America were of metal or a material other than glass. After that date, glass gradually became the most important medium for lighting devices, especially lamps. When the whaling industry was one of the most important in New England, the production of free-blown and pressed whale oil lamps by New England factories was a mainstay of their business. Pressing made glass lamps cheap and therefore available to the general user. Simple whale oil lamps burned more brightly with glass chimneys although not all used them, while the improved Argand burners required both chimney and a glass shade to diffuse their light—at ten candlepower, too harsh by the standards of the day. Count Rumford, writing in 1811 said, "No decayed beauty ought ever to expose her face to the direct rays of an Argand lamp," and doctors warned of eyestrain for users. In spite of these strictures, it would require five Argand lamps to equal the light provided by one modern sixty-watt bulb.

The discovery that kerosene or "rock oil" could be refined from petroleum, the oil boom in the Middle West, and the rising price of whale oil shifted the source of fuel to the Midwest and, as kerosene lamps *required* glass chimneys in order to burn properly, a whole new glass industry was born. There were one hundred forty-one patents granted for lamp chimneys and shades between 1855 and 1873 alone, and most of them were issued to manufacturers in the West. Kerosene continued to be the major fuel, notwithstanding the introduction of gas for home lighting, until the invention of the incandescent electric bulb by Edison in 1879 signaled the end of all other forms of lighting. The first light bulb blanks were blown for Edison by Corning Glass Works in 1879, but electricity did not completely supersede kerosene in rural areas until the rural electrification program at the end of the Great Depression

WINDOW GLASS

The production of flat glass for windows has been important since Caspar Wistar started his factory in 1739. There were two methods in use in this country. In one, called crown glassmaking, the gaffer blew a flat circular sheet, called a crown or table, to be cut into small but very clear panes. The cylinder method, used by Wistar, in which a long cylinder was blown and then opened and flattened, was faster and made larger sheets possible, but the flattening process rippled the surfaces of the glass, making it less clear. However, for the last half of the nineteenth century, most window glass was made by this method, the workmen making longer and longer cylinders and the factories becoming larger and more efficient.

Labor continued to be a problem in the glass factories, although after the Civil War there was less reliance on imported European labor. Unions were organized in the 1860's and by the 1880's wages and prices in the entire glass industry were closely regulated by the unions of the Window Glass Blowers, the Flint Glass Workers, and the Glass Bottle Blowers, as well as the National Flint Glass and Lime Association, the American Window Glass Makers Association and the Chimney Manufacturers Association. Wages were higher than those abroad and production was less, so manufacturers charged higher prices and there was an increasing use of child and female labor to save money. Tariffs protected manufacturers to some extent from European competition.

2

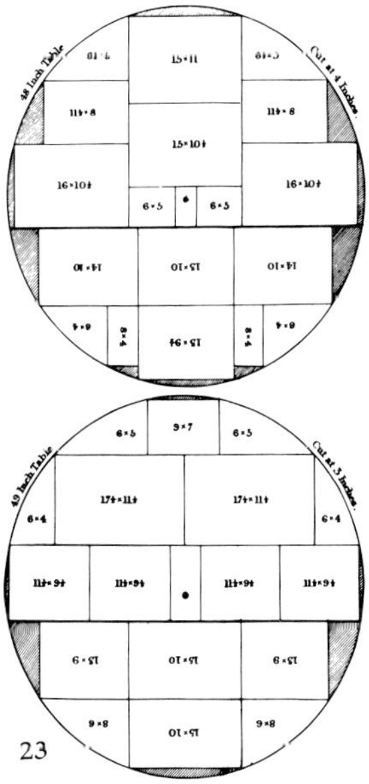

23

24

Fig. 22. *Cylinder glassmaking, Pittsburgh, 1884.*

Fig. 23. *Cutting a crown of glass into panes, from a glazier's manual, 1835.*

Fig. 24. *Pane sizes grew from the mid-eighteenth century (left) to the mid-nineteenth century (center) until by the 1890's full size panes were possible (right).*

Fig. 25. *Group of glas containers, ca. 1840–190*

GLASS CONTAINERS

Containers continued to be an important part of the glass industry and of American life. In 1880 more than twenty-five percent of the glass made here was for common bottles; slightly less than twenty-five percent was for windows and fifty percent was for tablewares and better quality containers.

Liquor flasks continued in popularity. The bitters and other patent medicines generated by the Temperance Movement were marketed in distinctive bottles which, in many cases, identified the product. Turlington's Balsam of Life, Swain's Panacea, Perry Davis' Pain Killer and Lydia Pinkham's Vegetable Compound are all well-known patent medicines which made fortunes for their originators and improved business for the bottle manufacturer—Hostetter's Celebrated Stomach Bitters, Drake's Plantation Bitters, Bininger's Bourbon Whisky Bitters and a host of others were polite substitutes for the liquor frowned on by the Temperance ladies.

Home preserving became popular in the second half of the century and foods were "canned" in glass jars made especially for that purpose. Both bottles and jars were produced by blowing into full-size molds and hand-finishing the necks and lips. Later in the century, less and less hand finishing was necessary; finally, in 1903, Michael Owens invented his automatic bottle blowing machine which signaled the end of an era. In spite of the inroads of plastic, aluminum and paper, glass is still preferred for many containers from cosmetics to wine.

LUXURY GLASS

As the Victorian idea of "good taste" changed after the Great Exhibition of 1851 in London and became more ornate, glass manufacturers sought new methods and styles of decoration for their products. Some glass of this period was imitative of other substances, such as mercury glass which imitated silver, or tortoise shell glass. Other glassmakers created new color effects. Glass could be made shaded or parti-colored by reheating portions of it, causing the reheated portions to "strike" or change color, and this became a popular method of decoration in the 1880's. Peachblow, Amberina, Burmese, Rose Amber, Agata, and Wild Rose were all varieties of shaded wares. Glass could also be enameled in imitation of porcelain (Royal Flemish, Crown Milano), beaded in imitation of coral (coralene) or have applied decoration in the Japanese style (Mat-su-no-ke).

These new styles fostered a desire to decorate every inch of the surface of a given piece and, as a result, patterns in glass cutting became more and more elaborate. The cut glass Mary Lincoln ordered for the White House in 1861 was simply cut and enlivened by the Presidential seal. The set made by the same manufacturer for the Philadelphia Centennial in 1876 was so profusely cut that not a half inch of surface was without decoration. This "brilliant" cut glass was the preferred wedding present, according to contemporary advertisements, from 1890 until 1914, and the two largest centers for its production were Toledo, Ohio, where the gigantic Libbey Company was located, and Corning, New York, where several smaller firms competed. World War I made raw materials hard to get and so increased prices that after the war cut glass was uneconomical to manufacture as well as out of fashion.

Fig. 26. *Shaded glasswares, l. to r. Wild Rose; Hobbs, Brockunier Peachblow; Agata; Burmese; Mt. Washington Peachblow, 1886–1890.*

Fig. 27. *Cut glass of the brilliant period. Plate and ruby-cased wineglass, T. G. Hawkes & Co. Russian pattern; goblet from President Harrison's White House set and swirl goblet, both C. Dorflinger and Sons. All 1880–1910.*

Fig. 28. *Favrile vase, Tiffany Furnaces, Corona, Long Island, ca. 1912. (62.4.61).*

ART NOUVEAU

L'art nouveau was an international artistic style embracing architecture and painting as well as the decorative arts. It began in France and was part of the art world from the 1870's until World War I. Emile Gallé of Nancy, France, was the first Art Nouveau glassmaker and his most prominent American counterparts were Louis C. Tiffany and Frederick Carder.

Tiffany was originally a decorator who became so fascinated with glass design that he opened his own factory to develop his ideas. He imitated the iridescence found on long-buried ancient glass; he tried forms from nature, Japanese styles, paperweight techniques, and generally made his glass as different as possible from the pressed and cut glass then fashionable. He was a leader in this movement until the 1920's.

In Corning, in 1903, Frederick Carder, an Englishman, set up the Steuben Glass Works, financed by T. G. Hawkes & Co., a local glass cutting firm. Carder made all types of colored and fancy glasswares, some similar to Tiffany's and others quite different. Other Tiffany competitors were the Union Glass Works in Somerville, Massachusetts, producing a line called Kew-Blas, Victor Durand's factory in Vineland, New Jersey, and the Quezal Art Glass and Decorating Company in Brooklyn. These and other factories turned out quantities of iridescent and colored wares sometimes true to the ideas of *l'art nouveau*, but often not. Carnival glass, a cheap pressed iridescent glass sometimes called "poor man's Tiffany" was made by the carload from 1900 until 1914.

After the War, simple, more geometric styles were in favor, instead of the flowing lines and curves of Art Nouveau. The glass of the 1920's reflects the austerity of post-war Europe as American styles followed those from abroad.

Out of the Depression, Steuben Glass rose again to prominence; limiting its material to colorless crystal and emphasizing copperwheel engraving, it raised the value of glass to that of a precious metal and every President since Harry S. Truman has given Steuben crystal as official gifts of State.

MODERN MASS PRODUCTION

Michael Owens, inventor of the first fully automatic bottle machine, had previously perfected a machine for blowing light bulbs, one in 1895 for blowing tumblers, and one for blowing lamp chimneys. These devices were the precursors of modern mass-production machinery. Tablewares began to be pressed by automatic machines in the 1920's, when the glassware now collected as "Depression Glass" was marketed. These were colorful table settings of machine-pressed glassware in pink, blue, green and other colors. Most of this glass was made in Midwestern factories and much of it was given away as premiums. The Depression was as devastating to the glass business as it was to most manufacturers and it was only by making such cheap wares that many factories stayed in business.

In 1926 the Wellsboro, Pennsylvania, plant of Corning Glass Works installed a ribbon machine, an automatic blowing machine used primarily to manufacture light bulbs and, after 1939, Christmas tree ornaments. It proved so productive (2,000 bulbs a minute, millions in a day) that the process was soon licensed to other glass companies and revolutionized that portion of the business.

The United States has continued its tradition of technological innovation. Today, other types of machines operate at tremendous speeds turning out tableware and bottles for such companies as Owens-Illinois; cook ware, laboratory ware for Corning Glass Works; tumblers for Anchor-Hocking, fruit jars for Ball Brothers, and window glass for Libbey-Owens-Ford and PPG Industries.

29

30

Fig. 29. *American factory-made glasswares, ca. 1930–1975.*

Fig. 30. *Corning Glass Works ribbon machine in operation, Wellsboro, 1956.*

Fig. 31. *Vase, Dominick Labino, 1969.* (70.4.24).

Fig. 32. *"Foursquare," Harvey Littleton, 1975.* (75.4.54).

31

32

STUDIO GLASS

Because of the complicated technology required for its production, glass was made entirely in factories until recently. The revival of interest in handcrafts such as pottery, woodworking, and weaving did not spread to glass until 1962, when a group of pottery students, led by Harvey Littleton of the University of Wisconsin, attempted to build a small furnace during a workshop at The Toledo Museum of Art. Dominick Labino, a scientist from Johns Manville, designed their furnace. From these small beginnings sprang the Studio Glass movement which in twelve short years has spread over this country and much of Europe.

Within four years of the Toledo workshop, enough glass was being produced, mostly in colleges and art schools, for the Toledo Museum to have a competitive national exhibition which attracted a large number of entrants. The first Toledo Glass National of 1966 showed the work of forty-three artists. The second, in 1968, showed fifty-seven artists and after the third exhibit Toledo was forced to discontinue the show because the sheer size of the movement made organization impossible. The Studio Movement, while primarily concerned with the work of individual artists in their own workshops, has also influenced design in large factories and has created a whole new range of art objects for the collecting public. With this movement bringing glass to the level of a medium in the fine arts, our oldest industry has suddenly become our newest art.

33

Fig. 33. *Blown form, made at Venini Studios, Venice, under the direction of Marvin Lipofsky, 1972. (74.3.120).*

Fig. 34. *"Ravenna Grand Jury," Henry Halem, 1972. (73.4.51).*

34

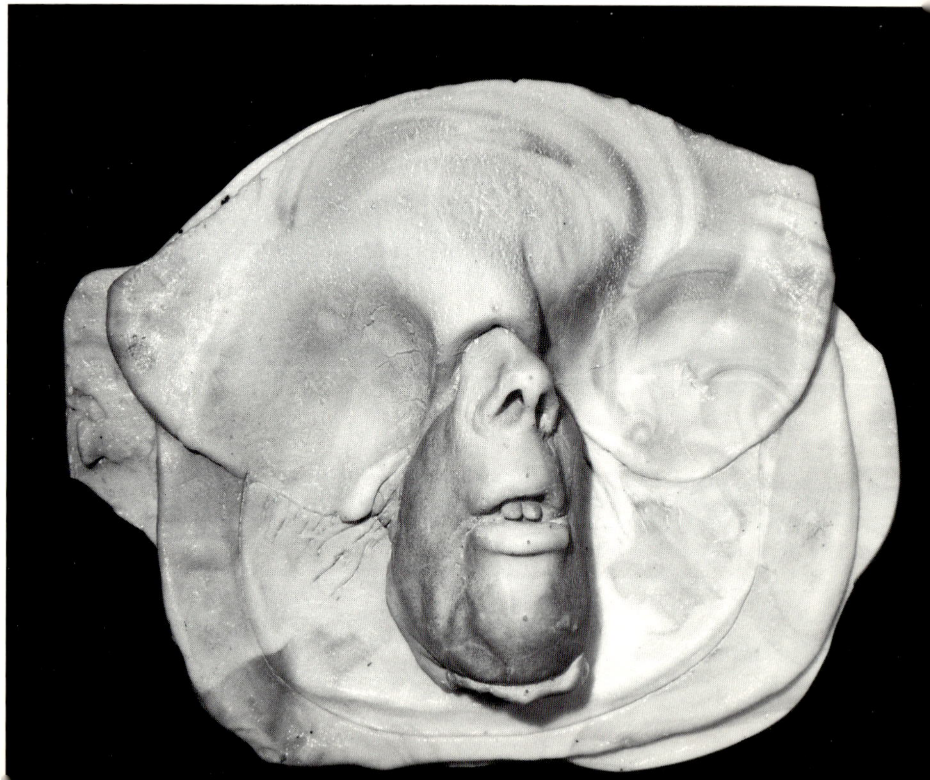

BIBLIOGRAPHY

Davis, Pearce. *The Development of the American Glass Industry.* Cambridge, 1949.

Gardner, Paul V. *The Glass of Frederick Carder.* New York, Crown Publishers, Inc., 1971.

Innes, Lowell. *Pittsburgh Glass—1797–1891. A History and a Guide for Collectors.* Boston (to be published), Houghton Mifflin, 1976.

Lee, Ruth Webb and Rose, James H. *American Glass Cup Plates.* Northboro, Mass., published by Ruth Webb Lee, 1948.

McKearin, George S. and Helen. *American Glass.* New York, Crown Publishers, Inc., 1950.

———. *Two Hundred Years of American Blown Glass.* New York, Crown Publishers, Inc., rev. ed., 1966.

McKearin, Helen. *Bottles, Flasks and Dr. Dyott.* New York, Crown Publishers, Inc., 1970.

———. *The Story of American Historical Flasks.* Corning, New York, The Corning Museum of Glass, 1953 (photomechanical reprint, 1968).

Neal, L. W. and D. B. *Pressed Glass Salt Dishes of the Lacy Period 1825–1850.* Philadelphia, published by the authors, 1962.

Revi, A. Christian. *American Cut and Engraved Glass.* New York, Thomas Nelson & Sons, 1965.

———. *American Pressed Glass and Figure Bottles.* New York, Thomas Nelson & Sons, 1964.

———. *Nineteenth Century Glass—Its Genesis and Development.* New York, Thomas Nelson & Sons, rev. ed., 1968.

———. *American Art Nouveau Glass.* Camden, New Jersey, Thomas Nelson & Sons, 1967.

Rose, James H. *The Story of American Pressed Glass of the Lacy Period 1825–1850. A Special Exhibition.* Corning, New York, The Corning Museum of Glass, 1954 (photomechanical reprint, 1967).

Wilson, Kenneth M. *New England Glass & Glassmaking.* New York, Thomas Y. Crowell Company, 1972.

Bu
PENN
Kent
Pittsburgh
INDIANA
OHIO
Wheeling
Cincinnati
VIRGINIA
Louisville
KENTUCKY